A FRENZY OF FLEAS

BY REBECCA STORM

CONTENTS

WHAT IS A FLEA?	4
PARTS OF A FLEA	6
LIVING IN HAIR	8
SUCKING BLOOD	10
SPREADING DISEASES	12
FINDING A HOME	14
INCREDIBLE JUMPERS	16
LAYING EGGS	18
FINDING A HOST	20
HUMANS AND FLEAS	22
TYPES OF PARASITES	24
BLOOD-SUCKING CREATURES	26
FUN FLEA FACTS	28
GLOSSARY	30
INDEX	32

Copyright © 2025 Hungry Tomato Ltd

First published in 2025 by Hungry Tomato Ltd
F15, Old Bakery Studios, Blewetts Wharf, Malpas Road, Truro, Cornwall, TR1 1QH, UK.

No part of this publication may be reproduced, stored in a retrieval system, or transmitted in any form or by any means, electronic, mechanical, photocopying, recording, or otherwise, without prior written permission of the copyright owner.

A CIP catalogue record for this book is available from the British Library.

ISBN 9781835694190

Printed in China

Discover more at
www.hungrytomato.com

Picture credits:
Abbreviations: m-middle, t-top, l-left, r-right, bg-background.

Science Photo Library: 10bl, 11b, Dr.PMarazzi 11tr, Martyn F.Chillmaid 13br, 18b, 19t, 20b, 21b, 22b, 24mr, 24bl, 25m, 25bl. Shutterstock: FC, 1bg, 6b, 7bl, 9b, 16b, 26bl, 27bl, 29br, 32bl; Frances Can Der Merwe 19br; ID1974 17t; Illustrissima 25b; Lightfield Studios 29mr; Lukas Jonaitis 7tr; MakroBetz 21tr; Marting Gstoekl 26mr; MC Jaarsveld 28mr; Millet Studio 15bg; Photowind 4b, 5br; Pitiya Phinjongsakundit 27tr; Sinhuyu Photographer 3bl, 14mr; Tomasz Klejdysz 12b, 28bl.

Every effort has been made to trace the copyright holders, and we apologise in advance for any unintentional omissions. We would be pleased to insert the appropriate acknowledgements in any subsequent edition of this publication.

DISCLAIMER:
Insects are fascinating, but best to stay away! Don't touch or handle them – some insects can sting or get aggressive when they feel threatened.

Words in **BOLD** can be found in the glossary.

WHAT IS A FLEA?

Fleas are small, wingless insects. They are so small, they are hard to see. Fleas can jump amazing distances… and they bite!

HOW DO THEY LIVE?

Fleas feed on blood, but they are not **predators**. Predators kill their **prey** before feeding, whereas fleas take blood from living animals! Creatures which feed off the living bodies of other creatures are known as **parasites**.

WHERE DO THEY LIVE?

Fleas live on the skin of **mammals** and birds, and are found wherever these animals live in the world. Animals that have parasites like fleas living on their bodies are known as **hosts**. Animals, such as bats and **rodents,** have the most, while horses are almost flea-free.

IT'S A BUGS WORLD

Insects belong to a group of animals known as **arthropods**. Adult arthropods have jointed legs, but do not have an inner **skeleton** made of bones. Instead, they have a tough outer "skin" called an **exoskeleton**. Most insects have at least one pair of wings.

Fleas have six legs but no wings, and they belong to the insect group of arthropods.

PARTS OF A FLEA

Fleas are only 3 mm (0.1 inch) long. They have six legs and no wings. Fleas have the same type of body as all other adult insects, which have three parts – **head**, **thorax**, and **abdomen**.

The body of a flea is very narrow and flat. They have lots of spines that point backwards, and bristles that stick out from their body.

The abdomen is the largest part of the flea's body. It contains the flea's **digestive system**.

The thorax is the middle part of the body, and the legs are attached here. The flea's back legs are longer and stronger than its front and middle legs.

HOW MANY LEGS?

Fleas and other insects are sometimes called "hexapods" because they all have six legs ("hex" means "six" in Greek). This can be a bit confusing – all insects are hexapods, but not all hexapods are insects! Some other bugs, such as **springtails**, have six legs, but they are not true insects.

Springtails can't bite or sting.

A flea's head has **antennae**, a brain, and a mouth. Most fleas also have eyes, but they have very poor sight. Some fleas do not have eyes at all.

LIVING IN HAIR

A flea has tiny, sharp claws on the ends of its feet so that it can get a good grip on the soft skin of its host. A flea's flat body makes it easy for it to move between the thousands of hairs or feathers that cover the host's skin.

Mammals and birds do not like having fleas! Most mammals constantly groom or scratch to try and get rid of parasites such as fleas. Birds clean their feathers for the same reason.

A flea moving through animal fur

When animals try to scratch away fleas, the bristles and spines on a flea's body become really useful! If a flea is pushed backwards, their spines and bristles catch on the host's hairs or feathers, stopping the insect from falling off.

ON THE OUTSIDE

Animal parasites can be divided into two groups. Fleas, and other similar bugs, are known as ectoparasites because they live on the outer surface of the host. Other bugs, such as tapeworms, live inside the host's body and are known as endoparasites.

A close up of an ectoparasite

SUCKING BLOOD

A flea's mouth is perfect for drinking blood! Each flea has three needle-like **stylets**. Two stylets pierce the host's skin, and the third allows the flea to suck up warm blood.

When the flea drinks, its abdomen fills up with blood. Fleas prefer to eat once a day but, if needed, they can go for several months without feeding.

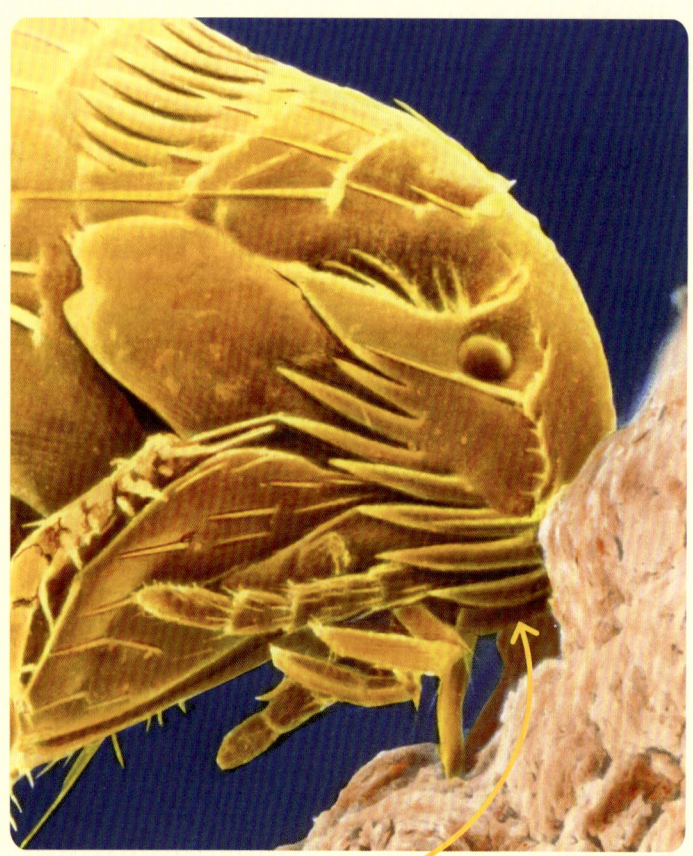

A flea sucking blood from a host

BARBED SECURITY

Flea stylets have rows of tiny, upward-pointing **barbs**. If the flea is detected while it is drinking, these barbs dig into the host's skin and make the flea very difficult to get off. Some fleas can stay attached to their host's skin for many days, holding on with these barbs.

A flea bite is almost painless, and most animals do not notice that they have been bitten at first. After a few minutes, the host's skin starts to react to substances on the flea's stylets – the bite begins to itch, and the host starts to scratch. But by this time, the flea has fed and moved on.

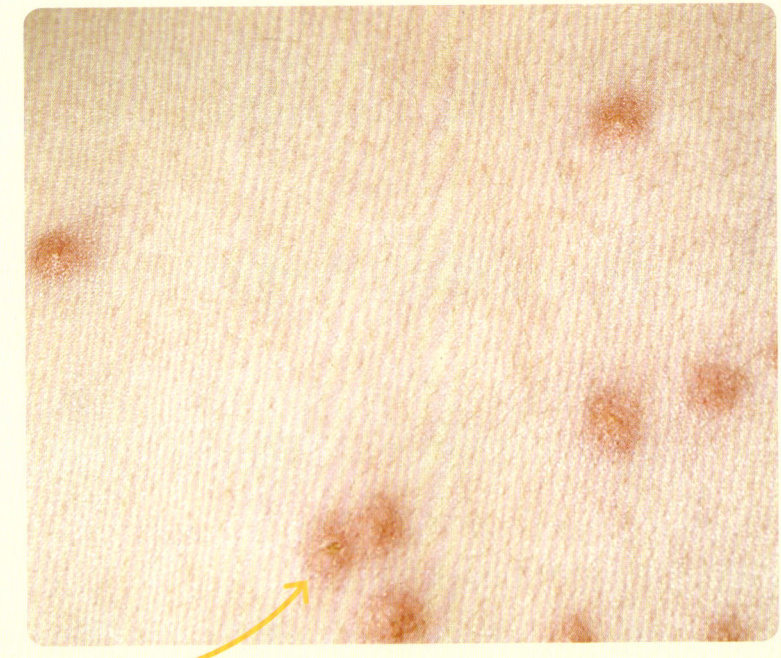

flea bites

The mouth parts of a flea, with the blood-sucking stylet in the middle

SPREADING DISEASES

Unfortunately, a flea is much more than just an itchy nuisance. It can also spread deadly diseases!

A single fleabite does not cause much harm by itself – the host only loses a drop of blood and itches slightly. The real problem is with what the flea had eaten just before...

When a flea sucks blood from a host, small traces of blood stay in its stylets. If the flea then moves to another host, those traces of blood get injected into the next host's body.

It is these small traces of blood that spread disease. If the first host has a disease, its blood will contain lots of disease-causing **microbes**. Even a tiny amount of blood left in a flea's stylets is enough to carry these microbes into the second host. When this happens over and over again, diseases can spread very quickly.

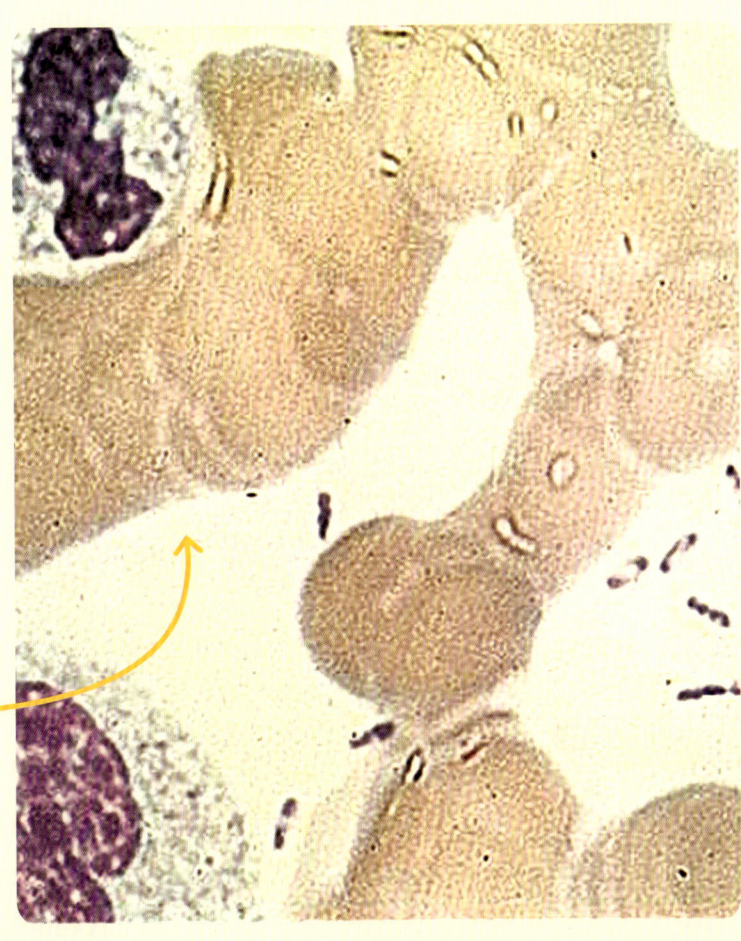

Microscopic view of microbes in a drop of blood

RABBIT KILLER

In many parts of the world, rabbits are serious pests because they eat lots of crops. In some areas, rabbit fleas have been used to reduce the number of rabbits. A few rabbits were infected with a disease called **myxomatosis**. Rabbit fleas quickly passed the disease from rabbit to rabbit, helping to keep their numbers low and the crops growing.

This rabbit has red, swollen eyes, which is a sign of myxomatosis.

FINDING A HOME

When fleas can't find a host to live on, they can go long periods without feeding at all. They will wait weeks, or even months, for the right host to come along.

FUSSY FLEAS

Fleas are often very choosy about where they live. Some fleas, however, are much less fussy. Cat fleas, for example, are quite happy to feed on the blood of other **species**.

If a host dies, its fleas are forced to look for a new home. If the host has been living in crowded conditions, such as a bat **roost** or a rabbit warren, then the fleas don't have to travel far.

A bat roost like this one would mean fleas have lots of hosts to choose from!

If the host has spent lots of time away from its home, the fleas may have a long wait before another host comes close enough to jump on.

INCREDIBLE JUMPERS

A flea spends most of its time either staying still or walking slowly over the skin looking for the right place to drink blood from. But, when it needs to, a flea can move further and faster than most living creatures!

When escaping from its host, or hopping onto a new host, a flea can jump about 200 times its own body length. That's the same as a human jumping over a 40-floor skyscraper in a single leap!

When they jump, fleas can **accelerate** more than 20 times faster than a plane!

STORED POWER

The secret of this jumping ability is in the muscles attached to the flea's back legs. When these muscles are triggered, they release their stored energy in one powerful burst, and push the flea high into the air!

A flea's powerful back legs help them jump.

LAYING EGGS

Male and female fleas can only mate and lay eggs if they have recently drunk fresh blood. After mating, the female flea can lay eggs, as long as there is a host available with a supply of blood she can drink from every day.

Female fleas lay some of their eggs while they are feeding, and these are spread all over the host's skin.

The eggs soon hatch into tiny **larvae** that are unable to drink blood like adult fleas. Instead, the larvae get their taste for blood by eating the droppings of the adults!

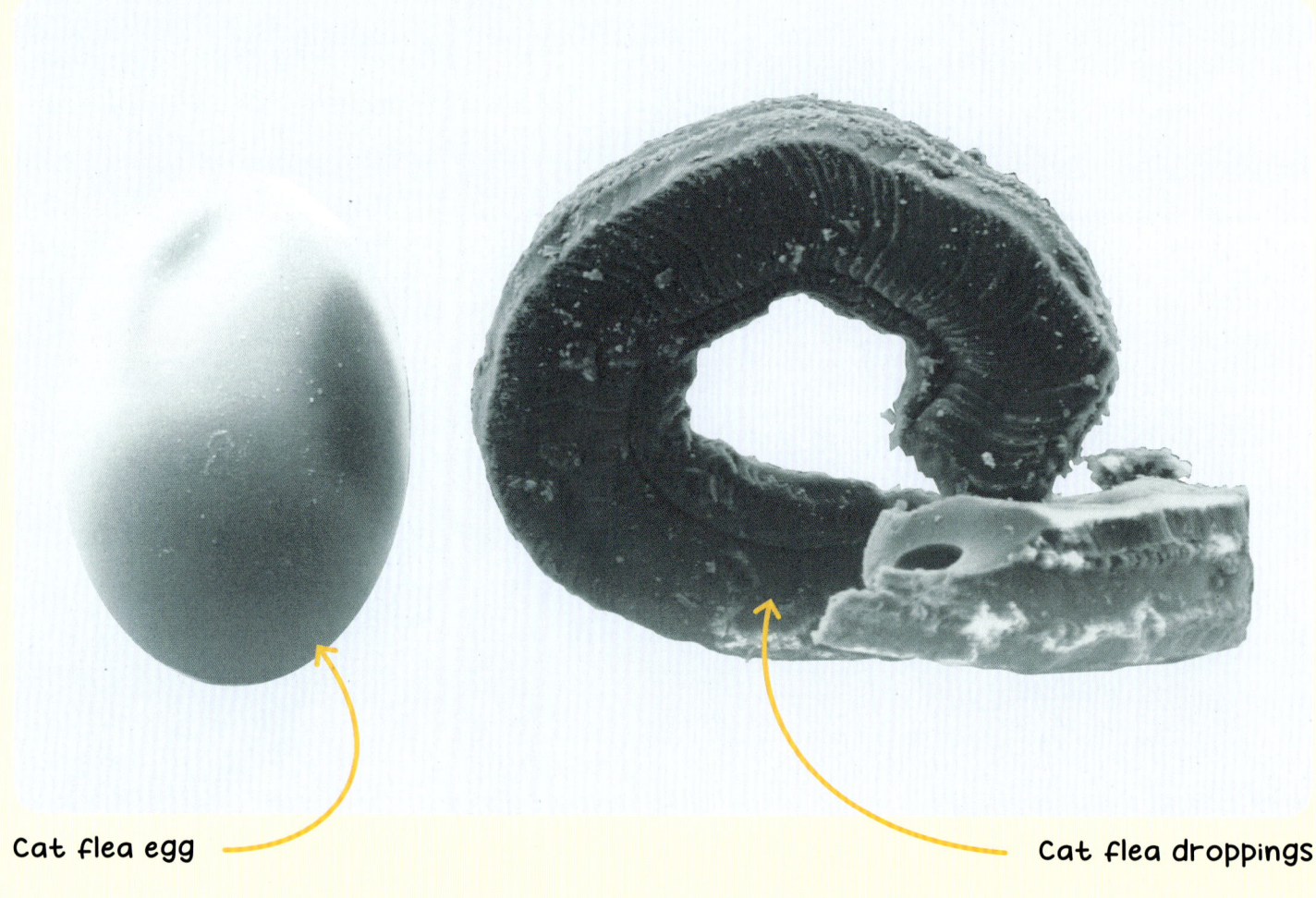

Cat flea egg

Cat flea droppings

In less than a month, the larvae are fully grown and they **pupate**. Flea larvae spin themselves into tiny **cocoons**. Inside their cocoons, the flea **pupae** wait patiently for the right moment before emerging as adult fleas.

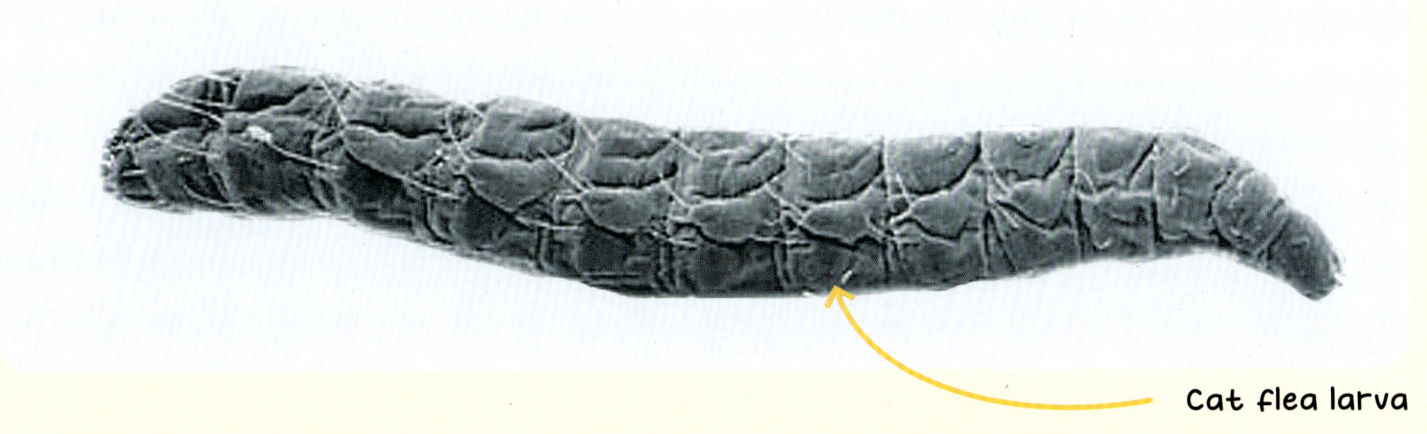

Cat flea larva

INSECT DEVELOPMENT

Insects develop from eggs in two different ways.

With many kinds of insect, including fleas, the eggs hatch into larvae that look very different from the adults.

However, with many other kinds of insect, such as cockroaches and grasshoppers, the eggs hatch into **nymphs** that already have the adult body shape.

A female aphid with her baby nymph

FINDING A HOST

A flea's pupation can take between six days to six months before it emerges from its cocoon. The flea waits until a suitable host is nearby, before it hops aboard!

When the pupa is in its cocoon, it cannot see, so it uses its other senses to tell when a host is nearby.

Cat flea cocoon

A pupa senses its surroundings through temperature and **vibrations** caused by movement. Pupae detect these vibrations through tiny bristles on their legs. The vibrations caused by an animal walking nearby are enough to let pupae know they can come out of their cocoon.

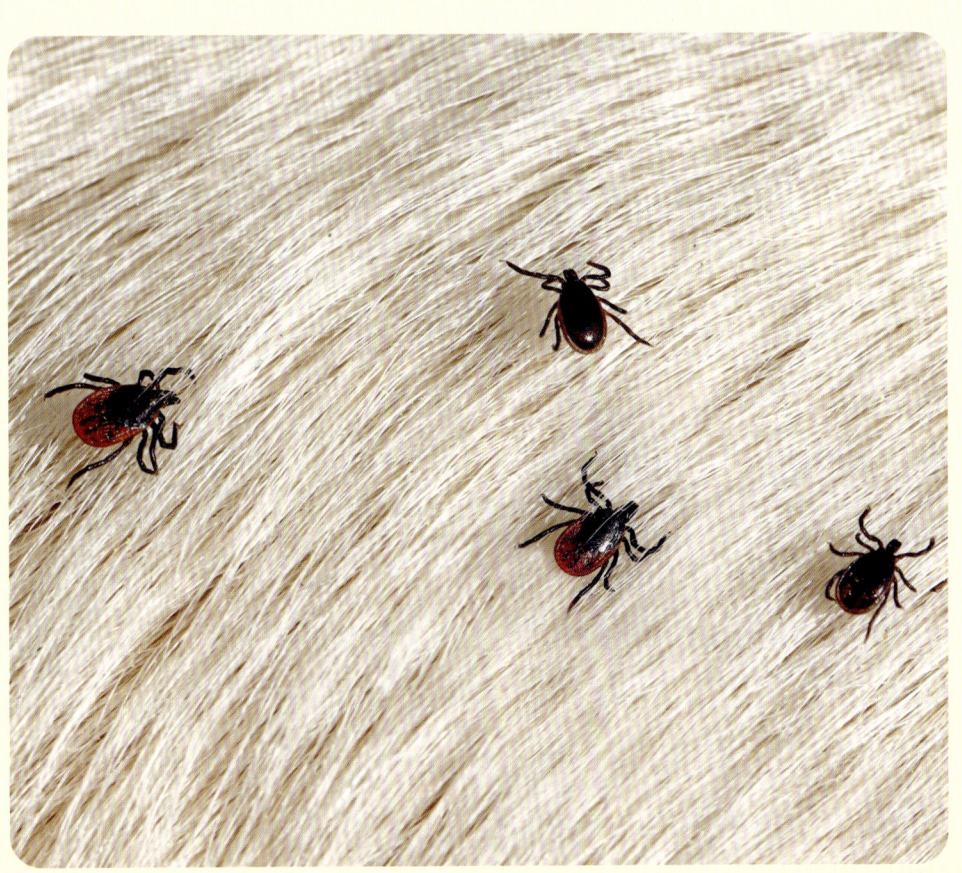

TOO HOT? TOO COLD?

Fleas are also very sensitive to temperature. They can find a suitable host in complete darkness, simply by sensing its body heat. The same is true for pupae. When a suitable animal comes close enough for a pupa to detect its body heat, the pupation stage ends immediately. A hungry flea emerges and leaps straight onto the nearest source of food!

HUMANS AND FLEAS

It is not surprising that human beings, who are the most widespread of all mammals on planet Earth, are hosts to their very own species of flea.

Human fleas are found wherever people are living. Fleas aren't picky; they will infest all sorts of places in their search for a host. A few hundred years ago, fleas were an uncomfortable part of daily life – lots of people had fleas.

A fleabite swells, itches for a few days, and then disappears. The bite itself does no real harm, unless the flea is carrying disease. Fleabites can spread dangerous human diseases, such as **typhus** and **bubonic plague**. These diseases were very common in the past, but not so much anymore.

BLACK DEATH

In the 14th Century, the Black Death (an outbreak of bubonic plague) wiped out one-third of Europe's population. Rat fleas carried the microbes which cause bubonic plague, and in the filthy conditions of the time, the fleas easily spread the disease from rats to humans.

People thought masks would keep themselves protected from the plague

TYPES OF PARASITES

Parasites normally all behave in the same way – but there are a few confusing exceptions!

CHIGOE FLEA

A pregnant female chigoe flea burrows into the skin of the host, usually in the feet. The female lives inside the host until she is ready to lay eggs, which causes intense itching for the host. A chigoe flea is sometimes called a "jigger".

CHIGGER

A chigger burrows into its host's feet, but it is not a flea – or an insect! It is the larva stage of certain mites that lay eggs on the ground. The eggs hatch into larvae that bite through skin and bury themselves in the feet of animals. They can itch even worse than jiggers, and both can pass on diseases.

LICE

Lice like to live on the skin of other animals. Some lice suck blood, while others do not. Bird lice, for example, usually feed on feathers. A few lice can't do much harm, but a group of lots of them can cause birds to lose their feathers!

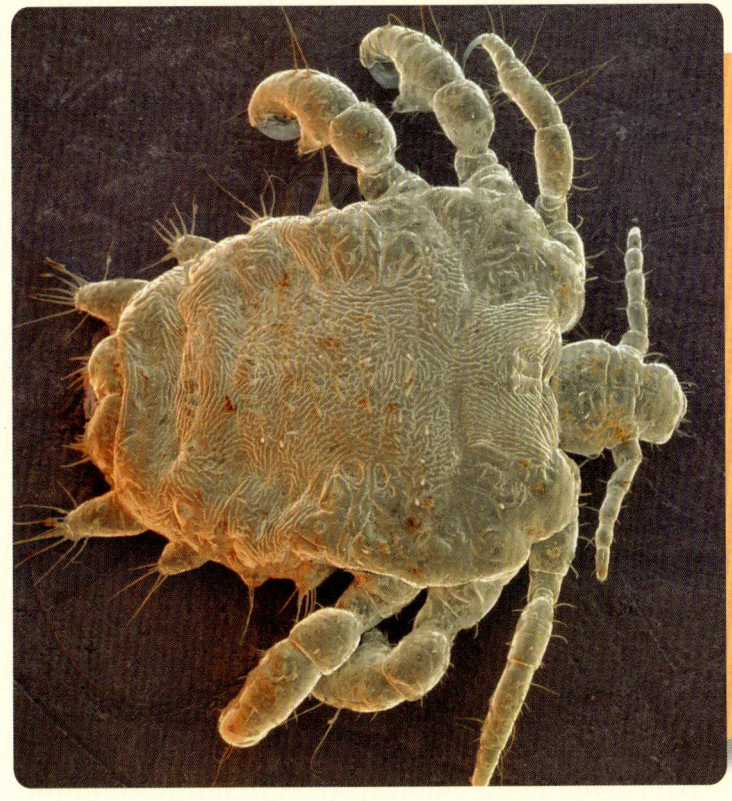

CRAB LOUSE

The crab louse likes to live in human body hair. They have strong claws for gripping on tight, and females produce special glue for sticking their eggs in place. Both young and adult crab lice feed on human blood, but they are not known to carry diseases.

BLOOD-SUCKING CREATURES

There are many other bugs that live by sucking blood.

MITES

Mites are **arachnids**, and are related to spiders. Some mites feed on plants, while others are parasites on animals of all sizes. The Varroa mite is a major problem for beekeepers because they lays eggs inside honybee colonies. When the young mites hatch, they feed on the honeybee larvae!

TICKS

Ticks are small arachnids that live in a similar way to fleas. All ticks are parasites that attach themselves to a host in order to suck blood. Like fleas, ticks attach themselves firmly to the host's skin. Some ticks can cause serious illnesses, such as Lyme disease.

MOSQUITOS

Mosquitos are small flying insects that, like fleas, have to feed on warm blood in order to breed. Mosquitoes, however, have a very different lifestyle. Their eggs are laid in water, and the larvae hatch out as water animals. Some mosquitoes carry deadly diseases such as malaria and yellow fever.

LEECHES

Leeches spend most of their time in water, and feed by sucking blood from fish, and other animals that live or feed near water. Leeches were once used in medicine to remove blood from patients. A special type of leech is still used in medicine today.

FUN FLEA FACTS

There's so much more to know about fleas! Delve into some fantastic facts about these tiny, blood-sucking creatures.

A FEMALE FLEA CAN...
lay about 2,000 eggs in her lifetime!

THE MOST COMMON FLEA...
is the cat flea.

A FEMALE FLEA CONSUMES...
15 times her own body weight in blood daily.

THE FLEA LIFE CYCLE CAN...
be completed in as little as 16 days.

ONCE A PET GETS FLEAS...
its surroundings must be completely cleaned, or else the pet will become re-infested!

IF A FLEA...
falls from its host, its tough exoskeleton prevents it from being injured.

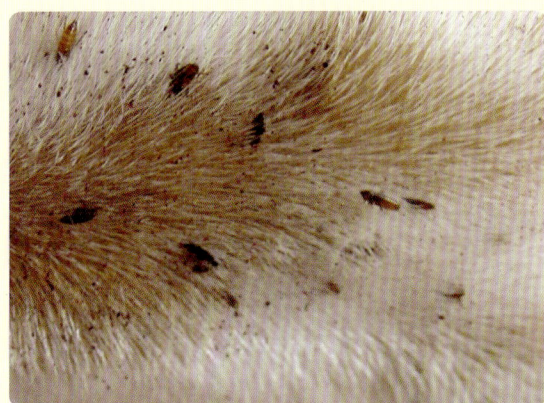

A DOG FLEA CAN...
live and breed on a dog for more than 100 days.

FLEAS HAVE..
been around for 100 million years!

THERE ARE OVER...
2,000 species of flea.

GLOSSARY

Abdomen – the largest part of an insect's three-part body; the abdomen contains most of the important organs.

Accelerate – to increase in speed.

Antennae – a pair of special sense organs found at the front of the head on most insects.

Arachnids – a group of bugs, with four pairs of legs and a two-part body, such as spiders, mites and ticks.

Arthropods – bugs that have jointed legs; insects and spiders are arthropods.

Barbs – sharp points facing the opposite direction on a weapon or tool that stops it being taken out of the victim.

Bubonic plague – a disease that normally affects wild rodents (see right), but which can be passed onto humans through fleabites.

Bug – one of a large number of small land animals that do not have a skeleton.

Cocoons – protective coverings of silk, produced by insect larvae (see below) to protect their bodies while they transform into adults.

Digestive system – the organs that are used to process food.

Diseases – conditions that cause part of a living thing to no longer work properly.

Exoskeleton – a hard outer covering that protects and supports the bodies of some animals.

Hosts – animals that provide food and a home for parasites.

Insect – a type of very small animal with six legs, a body divided into three parts, and usually two pairs of wings.

Larvae – a wormlike creature that is at the juvenile (young) stage in the life cycle of many insects.

Lice – a group of small insect parasites often found on birds and mammals.

Mate – one of a pair of animals that live or have babies together.

Mammals – a group of warm-blooded animals that have an internal skeleton and which feed their young on milk.

Microbes – tiny living things, so small that they can only been seen through a powerful microscope.

Myxomatosis – a disease that is passed on by rabbit fleas.

Nymphs – the juvenile (young) stage in the life cycle of insects that do not produce larvae (see left).

Parasites – any living thing that lives or feeds on or in the body of another living thing.

Predators – animals that hunt and eat other animals.

Prey – an animal that is eaten by other animals.

Pupae – insect larva that are in the process of turning into an adult.

Pupate – the process by which insect larvae (see left) change their body shape to the adult form.

Rodents – a group of small mammals that includes rats and mice, but not rabbits or shrews.

Roost – a resting or sleeping place used by birds and bats.

Skeleton – an internal structure of bones that supports the bodies of large animals such as mammals, reptiles, and fish.

Species – a group of living things that share characteristics and features.

Springtails – six-legged bugs that are not true insects.

Stylets – the hollow, pointed parts of a flea's mouth that are used to pierce skin and suck blood.

Thorax – the middle part of an insect's body where the legs are attached.

Typhus – a deadly disease that can be passed on to humans through fleabites.

Vibration – a quivering motion.

INDEX

A
abdomen 6, 10, 30
acceleration 17
adults 5, 6, 18-19, 25, 30-31
antennae 7, 30
arachnids 26
arthropods 5, 30

B
barbs 10, 30
bats 5, 14-15, 31
bees 26
birds 5, 8, 25, 31,
bites 4, 11, 12, 23, 24, 30
Black Death 23, 30
blood 4, 10, 12-13, 16, 18, 25, 26-27, 28, 31
bloodsucking insects 26-27
bristles 6, 9
bubonic plague 23

C
cat fleas 14, 19, 20, 28
chiggers 24
chigoe fleas 24
cleanliness 22
cockroaches 5
cocoons 19, 20-21, 30
crab lice 25

D
digestive system 6, 30
diseases 12-13, 23, 24-25, 26-27, 30-31
dogs 29
droppings 18

E
ectoparasites 9, 30
eggs 18-19, 24-25, 26-27, 28
endoparasites 9, 30
energy storage 17
exoskeleton 5, 29, 30
eyes 7, 13

F
feet-bloodsuckers 24
female fleas 18-19, 28
fleas 4-5, 6-7, 8-9, 10-11, 12-13, 14-15, 16-17, 18-19, 20-21, 22-23, 24-25, 26-27, 28-29, 30-31
food 21, 30

G
grasshoppers 19

H
head 6-7, 30
headlice 25
hexapods 7
honeybees 26
horses 5
hosts 5, 8-9, 10-11, 12-13, 14-15, 16, 18, 20-21, 22, 24, 26, 29, 30

I
insects 4-5, 6-7, 19, 24, 27, 30-31

J
jiggers 24
jumping power 16-17

L
larvae 18-19, 24, 26-27, 30-31
legs 5, 6-7, 17, 21, 30-31
lice 25
life cycle 28

M
mammals 5, 8, 22, 31
mating 18-19
microbes 13, 23, 31
mites 26
mosquitoes 27
mouth parts 11
muscle power 16-17
myxomatosis 13, 31

N
nymphs 19, 31

P
parasites 4-5, 8-9, 26, 30-31
pets (see cat fleas, dogs)
plague 23
predators 4, 31
pupae 19, 21, 31
pupation 20-21,

R
rabbit fleas 13, 14, 31
rat fleas 23
rodents 5, 30-31
roosts 14-15, 31

S
senses 20-21
skeleton 5, 30-31
spines 6, 9
springtails 7
stick-tight fleas 11
storing energy 17
stylets 10-11, 12-13, 31

T
tapeworms 9
temperature 21
thorax 6, 31
ticks 26
typhus 23, 31

V
varroa mite 26
vibrations 21

W
wings 5, 6, 30